W9-COL-191

Trees

Tony Johnston · Illustrated by Tiffany Bozic

A Paula Wiseman Book · Simon & Schuster Books for Young Readers
New York London Toronto Sydney New Delhi

SIMON & SCHUSTER BOOKS FOR YOUNG READERS
An imprint of Simon & Schuster Children's Publishing Division
1230 Avenue of the Americas, New York, New York 10020

The text for this book was set in Lomba Book.
The illustrations for this book were rendered in acrylic paint on wood.
Manufactured in China · 0721 SCP · First Edition
2 4 6 8 10 9 7 5 3 1
Library of Congress Cataloging-in-Publication Data
Names: Johnston, Tony, 1942– author. | Bozic, Tiffany, illustrator.
Title: Trees / Tony Johnston ; illustrated by Tiffany Bozic.
Description: First edition. | New York : Simon & Schuster Books for Young Readers, [2021]
"A Paula Wiseman book." | Includes bibliographical references. | Audience: Ages 4–8. | Audience:
Grades 2–3. | Summary: "Part poetry, part celebration of nature, turn the page of this book and
enter the majestic world of trees"— Provided by publisher.
Identifiers: LCCN 2021000126 (print) | LCCN 2021000127 (ebook) |
ISBN 9781534475175 (hardcover) | ISBN 9781534475182 (ebook)
Subjects: LCSH: Trees—Juvenile literature.
Classification: LCC QK475.8 .J64 2021 (print) | LCC QK475.8 (ebook) | DDC
582.16—dc23
LC record available at https://lccn.loc.gov/2021000126
LC ebook record available at https://lccn.loc.gov/2021000127

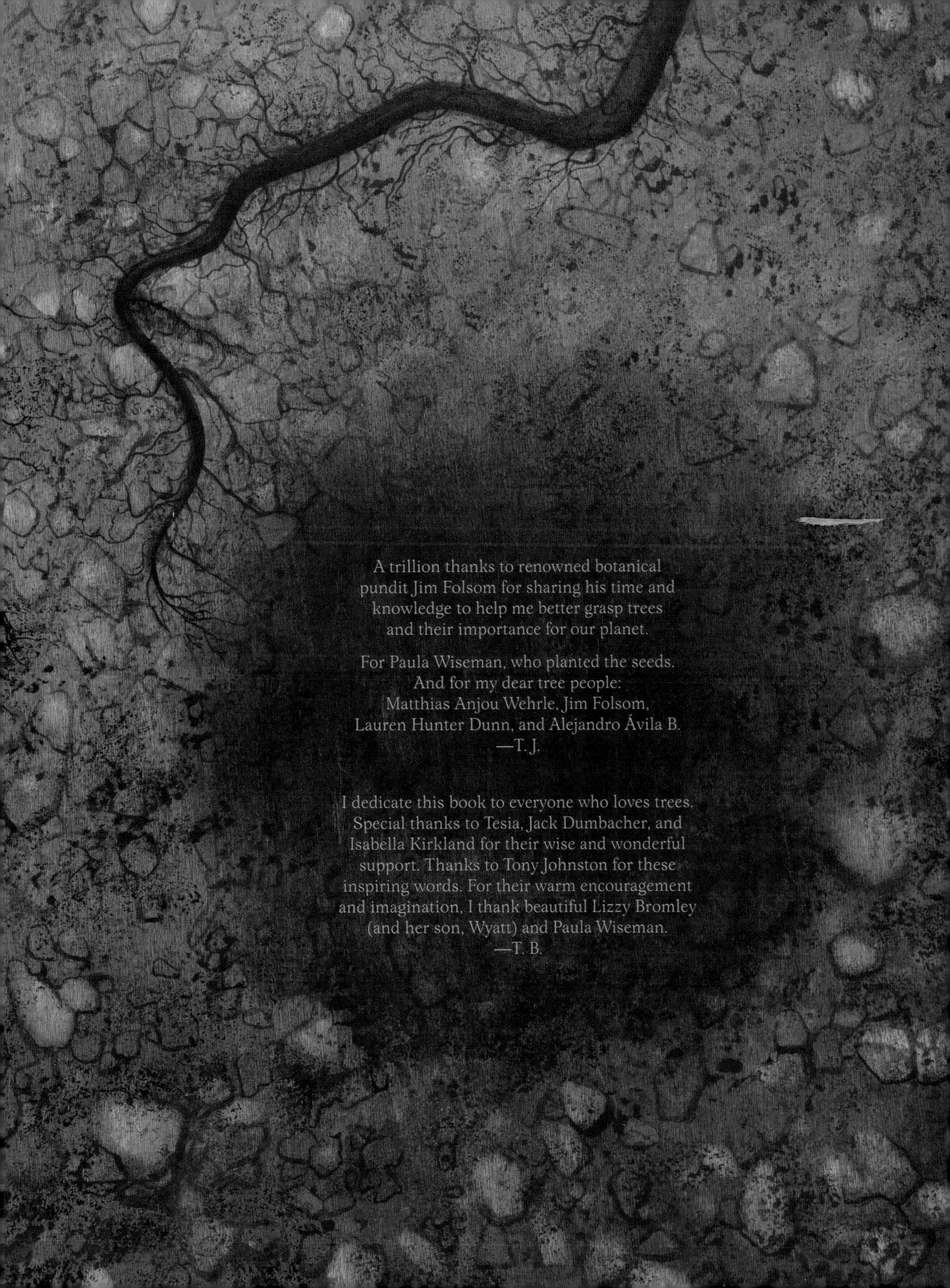

A trillion thanks to renowned botanical
pundit Jim Folsom for sharing his time and
knowledge to help me better grasp trees
and their importance for our planet.

For Paula Wiseman, who planted the seeds.
And for my dear tree people:
Matthias Anjou Wehrle, Jim Folsom,
Lauren Hunter Dunn, and Alejandro Ávila B.
—T. J.

I dedicate this book to everyone who loves trees.
Special thanks to Tesia, Jack Dumbacher, and
Isabella Kirkland for their wise and wonderful
support. Thanks to Tony Johnston for these
inspiring words. For their warm encouragement
and imagination, I thank beautiful Lizzy Bromley
(and her son, Wyatt) and Paula Wiseman.
—T. B.

Trees love sky.
They love all that blue
above them.

Trees love clouds.
They reach high
to touch them.

Days, they hold out their limbs
for songbirds to come.

Nights, they hold out their limbs
for stars. And moon.

Trees are beautiful.
In sun, their leaves shine.

In rain, they gleam.

Their good rough bark says,
I am strong.

Some trees bloom. Each year
their blossoms burst forth,
white like snow.

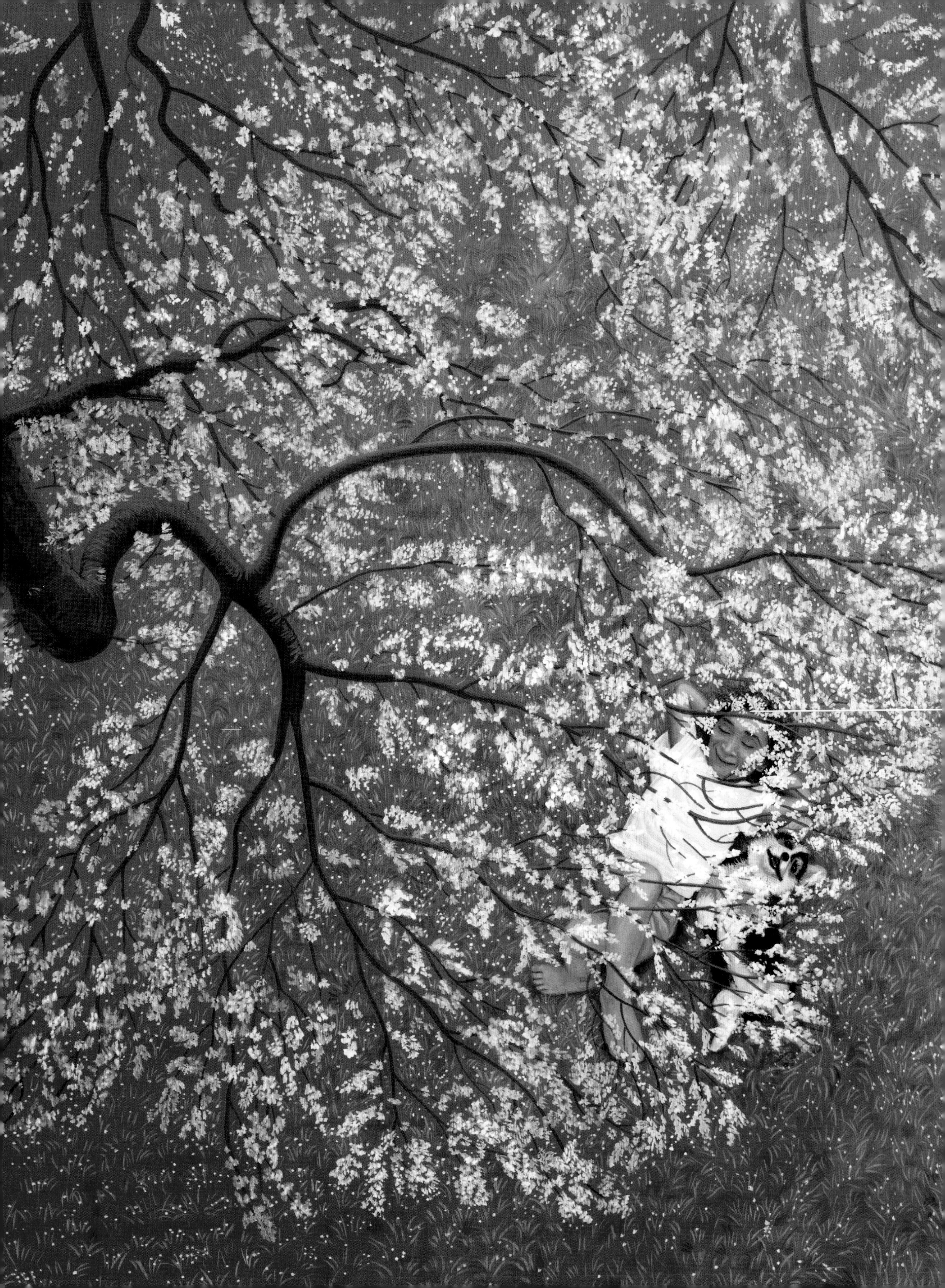

They make people happy
like snow.

Some trees are old, old, old.
Their ringed cores hold stories.
Of summers. Winters. Years.

Trees are friends.
Their branches say,
Come. Come. Climb.

So children do. Raccoons climb them too.
Like friends, trees stay.
They do not go away.

Trees give shade
to wanderers passing by.

I like to read in the shade of a tree.
Just a tree and me.

Author's Note

In addition to the books listed below, much of the author's note relies on conversations with Jim Folsom, Director of the Gardens at The Huntington in San Marino, California. You can learn more at huntington.org.

The ancient Greeks thought that trees were upside-down beings. They believed that the roots were their heads and so their most important part. They thought that trees ate dirt. After centuries of study, scientists know that trees are not animals at all. They belong to the kingdom of plants.

A tree has three main parts: roots, stems or branches, and leaves.

Half the life of a tree happens underground. The roots live in the world of dirt. They stabilize the whole tree and bring water and nutrients necessary for growth. They benefit from special relationships with fungi, through which additional nutrients are provided.

The branches and leaves live in the world of air. From the air, trees take carbon dioxide, a gas that at high levels is dangerous to humans. During photosynthesis, using carbon dioxide, water, and sunlight, trees give off oxygen, a gas necessary for human life.

Trees are "water-cycling machines." They help cool the earth. As water evaporates from them, they pull heat from the environment.

Trees are very complicated. Recent scientific studies show that they live in communities. Though not in a conscious way, they send "distress signals about drought and disease . . . or insect attacks." Other trees react to these signals. Apart from their root systems, they also communicate through the air, through scents.

Today our planet is in crisis. Among other things, the use of fossil fuels and cattle for food have contributed to global warming. The burning of fossil fuels increases the amount of carbon dioxide in the atmosphere. Clearing forests to graze cattle decreases the oxygen supply in the air, leaving a higher concentration of carbon dioxide.

One study shows that trees are "by far the cheapest climate-change solution." To combat climate change, many movements throughout the world encourage people to plant trees. The Trillion Trees organization is one of these.

In 2007, in Germany, nine-year-old student Felix Finkbeiner gave a school report known as Plant-for-the-Planet. His idea: to stop global warming, children of all countries could plant one million trees.

With various supporters, including the United Nations, Felix's dream has bloomed. Plant-for-the-Planet is now a worldwide movement. The Trillion Trees organization is part of it.

The Alfredo Harp Helú Foundation in Oaxaca, Mexico, is another group dedicated to reversing climate change. From its nurseries, which grow native trees, local people replant and care for the trees throughout the region. Its goal for 2020 was to plant five million trees in the state of Oaxaca.

Trees are many things. They are sources of life-giving oxygen and of hope in the fight against climate change. They are fantastic subjects for scientists to study. Still, when I look at a tree, I see a friend, a home for birds, a place to play. I see history, I see beauty, I see poetry.

Illustrator's Note

Where do you think the pages in this book come from? That's right, trees.

The pictures in this book were also made possible by trees: they were all painted on wooden panels.

For twenty years, I have created hundreds of paintings on maple because each panel is unique, just like each of us. Patterns in wood grain, like fingerprints, are shaped by sunshine, water, and nutrients. If you look closely, you will see that the wood grain shows through many of the paintings. Some even have the grain completely exposed.

My husband, daughter, and I live in an old cottage nestled under tall redwood trees in California. Each day we are surrounded by trees. I turn over rotting logs to find salamanders and banana slugs. I read in the shade of trees with my daughter. We travel to the Sierra Nevada mountains to see ancient pines. We can all thank trees for providing us with air to breathe, houses to live in, and books to read! Many other species depend entirely on trees to live.

I hope my paintings inspire you. Let's all celebrate the diversity, beauty, and importance of our connection to the natural world.

One of the best ways we can start is to go outside! Ask your friends and family to explore the woods and climb trees with you. Or you can take action to help preserve trees for future generations. Learn more about them. Plant native trees in your backyard. Join a group that's restoring trees in your area.

I encourage you to share all the amazing and wonderful things you discover in nature with others. Your curiosity is contagious! Remember, we protect what we *love*.

Suggestions for Further Reading

FOR CHILDREN

Udry, Janice May. *A Tree Is Nice*. New York: HarperCollins, 1987.

FOR ADULTS

Books

Preston, Richard. *The Wild Trees: A Story of Passion and Daring*. New York: Random House, 2007.

Wohlleben, Peter. *The Hidden Life of Trees: What They Feel, How They Communicate—Discoveries from a Secret World*. Vancouver: Greystone Books, 2016.

Articles

Borenstein, Seth. "Best Way to Fight Climate Change? Plant a Trillion Trees." *AP News*, July 4, 2019. https://apnews.com/article/8ac33686b64a4fbc991997a72683b1c5.

Plant-for-the-Planet. "Our Story." Aims and Vision. https://www.plant-for-the-planet.org/en/about-us/aims-and-vision#.

Robbins, Jim. "Chronicles of the Rings: What Trees Tell Us." *New York Times*, April 30, 2019. https://www.nytimes.com/2019/04/30/science/tree-rings-climate.html.

There are many organizations that are friends to trees and work to help preserve them and keep them healthy. Here is a partial list:

Alfredo Harp Helú Foundation: fahho.org
American Forests: americanforests.org
Arbor Day Foundation: arborday.org
Annie E. Casey Foundation: aecf.org
Eden Reforestation Projects: edenprojects.org
International Society of Arboriculture: isa-arbor.com
Nature Conservancy: nature.org
One Tree Planted: onetreeplanted.org
Plant-for-the-Planet: plant-for-the-planet.org
Trees for the Future: trees.org
Trillion Trees: trilliontrees.org
Wangari Maathai Foundation: wangarimaathai.org

A List of Trees in This Book

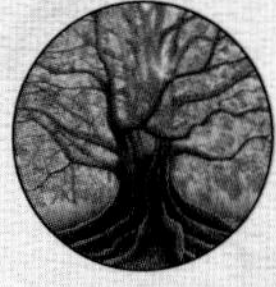
kapok tree (*Ceiba pentandra*)

redwood trees (Sequoioideae)

apple tree (*Malus domestica*)

assorted leaves: bay laurel, tan oak, redwood, acacia

coast live oak saplings (*Quercus agrifolia*)

Pacific madrone tree (*Arbutus menziesii*)

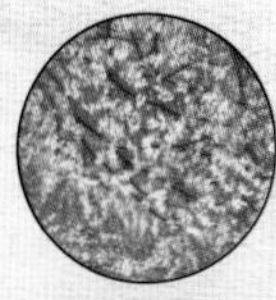
weeping cherry tree (*Prunus pendula*)

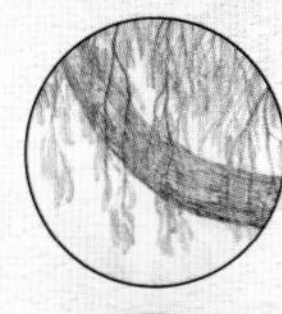
weeping willow tree (*Salix babylonica*)

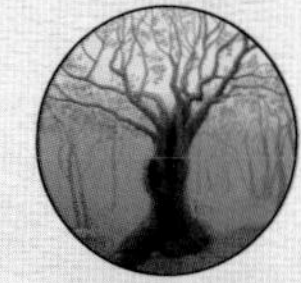
camphor tree (*Cinnamomum camphora*)

Japanese maple tree (*Acer palmatum*)

bristlecone pine tree (*Pinus longaeva*)

Jeffrey pine tree (*Pinus jeffreyi*)

roots: unidentified

Japanese emperor oak tree (*Quercus dentata*)

birch tree (*Betula*)

life cycle of redwood forest floor

maple tree (*Acer*)

Pacific madrone tree (*Arbutus menziesii*)

California pepper tree (*Schinus molle*)